Ilson Rodrigues dos Reis Junior
Amanda da Silva Mário

Sustainable design for flood drainage in the city of Araxá/MG

Ilson Rodrigues dos Reis Junior
Amanda da Silva Mário

Sustainable design for flood drainage in the city of Araxá/MG

New concept of urban planning

Imprint

Any brand names and product names mentioned in this book are subject to trademark, brand or patent protection and are trademarks or registered trademarks of their respective holders. The use of brand names, product names, common names, trade names, product descriptions etc. even without a particular marking in this work is in no way to be construed to mean that such names may be regarded as unrestricted in respect of trademark and brand protection legislation and could thus be used by anyone.

Cover image: www.ingimage.com

This book is a translation from the original published under ISBN 978-613-9-66633-1.

Publisher:
Sciencia Scripts
is a trademark of
Dodo Books Indian Ocean Ltd. and OmniScriptum S.R.L publishing group

120 High Road, East Finchley, London, N2 9ED, United Kingdom
Str. Armeneasca 28/1, office 1, Chisinau MD-2012, Republic of Moldova, Europe
Printed at: see last page
ISBN: 978-620-8-05246-1

AUTHORS' BIOGRAPHIES

Amanda da Silva Mário was born in Ribeirão Preto (SP) in 1995. Graduated in Civil Engineering from the Araxá Planalto University Centre in 2018.

Ilson Rodrigues dos Reis Júnior was born in Araxá (MG) in 1995. He is studying Civil Engineering at the Araxá Planalto University Centre in 2018.

Cinthia Aparecida Carneiro Lana was born in Araxá (MG) in 1988. She graduated in Architecture from PUC Minas in 2012 and has a Master's degree in Structural Engineering from UFMG in 2016.

Diogo Aristóteles Rodrigues Gonçalves was born in Araxá (MG) in 1986. He has a degree in Agronomy from FAZU - 2009-, a postgraduate degree in University Teaching from the Centro Universitário do Planalto de Araxá -2015-, a Master's degree in Plant Production from UFV/CRP - 2014- and is currently studying for a PhD in Agronomy at UFU/MG.

Summary

With the development of the city of Araxá, there have also been problems such as flooding in some parts of the city's lower streets, caused by a large amount of water that does not infiltrate the urbanised surface. Based on the problems generated by the accumulation of water in the area, such as difficulties for pedestrians and vehicles to move around and disruption to nearby establishments, there was a need to develop a preliminary project - to prevent flooding - that could be applied in such a way as to minimise or even extinguish the issue. To

To prevent these situations, the use of sustainable urban devices is being proposed to improve the environment, minimising both the impact of flooding and the impact on the environment. The format of the work was defined in such a way as to facilitate understanding of the subject and InfraWorks software was used to build the idealisation of the project in order to arrive at a forecast of how the area studied will look after the implementation of the drainage system. With this work, it is hoped that the community interested in the subject can use the material to help develop other projects or follow on from this one. The design uses sustainable drainage methods that favour the infiltration of water into the soil and reduce the speed of surface runoff and, consequently, the risk of flooding.

Keywords: Flooding. Drainage. Surface runoff. Flooding. Urban Planning. Project.

Summary

CHAPTER 1

Introduction

Urban planning, as described by Duarte (2012, p. 24), can be understood as "an instrument that seeks to study cities by predicting the changes that may occur in their structure. Urban layout, urbanism and urban management must be studied and, in addition, urban planning analyses physical, cultural and economic characteristics".

According to Souza (2002. p.46), managing refers to the present in order to administer something and planning to the future and the forecasting of processes. He explains that "planning is preparation for future management, seeking to avoid or minimise problems and expand margins for manoeuvre".

Botelho (2011) describes that a well-planned city has a positive impact on the economy as it can partner with neighbouring municipalities to boost its development, issues involving accessibility and safety are better discussed and the city gains greater credibility.

"However, cities are currently growing on a large scale, which leads to various problems that affect the population's quality of life. Due to this increase, masterplans need to be frequently revised" (DUARTE, 2012).

The consequences of a disorderly occupation of the urban environment are the removal of natural vegetation, the occupation of areas at risk of landslides, land use and occupation, an increase in waterproofed urban areas, an increase in runoff rates, which can generate erosive processes on slopes, silting up of rivers and flooding in urban areas. This is intensified by constant urban expansion (RODRIGUES, 2014).

These are modifying factors which, in many situations, result in an increase in the volume of surface runoff, causing watercourses to overflow (POMPÊO, 2000; TUCCI, 2003).

Floods are increasing in frequency and magnitude due to waterproofing, land occupation and the construction of rainwater networks.

According to Tucci (2003), as the city urbanises, the following impacts generally occur:

- Increase in maximum flows (by up to 7 times) and their frequency due to the increase in run-off capacity through conduits and channels and the sealing of surfaces;

- Increased sediment production due to unprotected surfaces and the production of solid waste (rubbish);

- Deterioration of surface and groundwater quality, due to street washing, transport of solid material, clandestine connections of cloacal and pluvial sewage, and contamination of aquifers; Due to the disorganised way in which urban infrastructure is implemented, such as:

- Deposition and obstruction of rivers, canals and conduits by rubbish and sediment;

- Inadequate drainage projects and works, with diameters that decrease downstream, drainage without drainage, among others.

In the city of Araxá, a Brazilian municipality in the state of Minas Gerais, located in the Alto Paranaíba region, the city's development has also led to problems related to flooding in some parts of the city's lower streets, caused by a large amount of water, coming from rain or from an accumulation, not sufficiently absorbed by the soil due to waterproofing or occupied areas.

The lack of planning for the execution of works that strongly address these floods and the financial limitations end up preventing good urban development, which leads to new implementations to put an end to the resulting floods in the city.

The volume of water runoff into the city's low-lying streets was identified and the main areas where flooding occurs in Araxá were mapped in order to develop a project to put an end to these problems related to water resources.

Flooding poses major health risks to the population and, in more serious cases, leaves residents homeless and destroys businesses.

Considering the importance of these factors, the creation of a project at this level helps to elucidate these conditions, creating a clear, effective and sustainable design using street furniture.

CHAPTER 2

History of Araxá

The city of Araxá is a Brazilian municipality in the state of Minas Gerais, located in the Alto Paranaíba region, with a population estimated by the IBGE in 2015 at 104,283 inhabitants. Its geomorphological conformation includes flat lands and hills with a relief that shows variations between cerrado and mountain ranges.The city's main economic source involves mining and also a large contribution from tourism, which allows Araxá to exploit its medicinal waters, manufacture soaps and skin creams as well as having one of the richest handicrafts in the region[1] . The layout of the city is made up of a closed linear grid with a central square where the street goes around the square, which in the city is the Igreja Matriz de São Domingos. Figure 1 shows a photo of the town from above.

Figure 1. City of Araxá
Source: Prepared by the authors.

---

[1] https://www.mfrural.com.br/mobile/cidade/araxa-mg.aspx

CHAPTER 3

Urban Planning

The term urban planning, which comes from Europe and the USA, emerged at the end of the 19th century and the beginning of the 20th century, due to the need to solve urban problems in cities as a result of industrialisation" (SOUZA, 2006, p21).

Population growth, urbanisation and the development of cities are factors that influence a better quality of life in the urban environment, but they also have impacts such as the reduction of natural resources, deforestation, the extinction of species and, above all, soil and water pollution. To try to minimise these effects, it is necessary to build well-planned cities that function efficiently, effectively and sustainably.

"Sustainable urban planning takes into account issues relating to the structural and organisational configurations of the city, such as location, sanitation, transport, public roads, demographic growth and so on. The way to achieve good urban planning is through each city's masterplan." (ACR ARQUITETURA, 2017).

According to Pinheiro (2012, p. 54 apud BRASIL, 1988, art. 182), the 1988 Constitution strengthened local power and delegated to municipalities the competence "which should aim to organise the full development of the social functions of the City and guarantee the well-being of its inhabitants".

According to the Araxá Institute for Planning and Sustainable Development (IPDSA), *the masterplan is the* instrument, *the* law, that establishes, on behalf of the community, *the* responsibilities and actions to be undertaken during a given period, capable of taking the city to a higher stage of development."

Also according to the Institute, the masterplan aims to propose actions that will transform the city, actions that must have the consent of the population and be implemented effectively.

Also, according to Miguel, Pedroso and Hoffmann (2009, p. 1), "Urban planning is the organisation of urban spaces in a specific way, that is, from urban planning to the current Master Plans" (apud Villaça, 1999). It can therefore be concluded that the masterplan is a tool that seeks to plan the urban environment in order to improve the lives of the population.

3.1 Stages of urban planning

According to Duarte (2012, p. 29) "planning is a process whose result, always partial, is the plan. The plan has parts; planning has stages". The following stages have been described in the book Urban Planning by Fábio Duarte.

■ Diagnosis

This first stage can be referred to as an analysis of the situation by means of an inventory, which is an integral part of the diagnosis stage, where data will be collected and organised on a given geographical area.

Prognosis

Once the diagnosis has been made and the city's situation has been analysed, we start to think about what the city will look like tomorrow. This stage will serve to seek solutions and alternatives to the situations that may occur in the area studied.

■ Proposals

"Proposals are the result of an urban planning process and are what transform a foreseeable future into a possible future." (DUARTE, 2012, p. 34) At this stage of the urban planning process, alternative ways are created for the population to participate in the day-to-day running of the city, since these proposals present infrastructure works that will help improve the development of the city and the quality of life of its inhabitants.

■ Urban Management

This last stage, which is just as important as the others, responds to the needs and demands of the population, ensuring that the proposed plans are implemented effectively and on schedule.

3.2 Proper urban planning

"Urban centres are increasingly agglomerated with people and real estate, so they must be accessible to their inhabitants. As transformations occur in the social, political and economic scenarios of cities, increasingly complex forms of spatial organisation are generated, which expand to metropolitan, regional and national levels" (VIDAL, 2012, p. 4). Proper planning involves establishing a relationship between urban services such as treated water, electricity, adequate and high-quality means of transport, well-dimensioned infrastructures, among others, and presenting projects that are economical and sustainable. Figure 2 below shows the city of Brasilia, an example of proper planning.

Figure 2: Urban planning in Brasilia
Source: SINFOR, 2016.

However, this is not offered in the right way. Cities that don't have good urban planning tend to be poorly planned, poorly structured and poorly lived in. It is the municipality's duty to make these adjustments to city infrastructure. According to Honda et al (2015, p. 64): The Federal Constitution of 1988 (Brazil, 1988), in its chapter on urban policy, assigned to the municipality the functions of control, planning, management and urban development. This shows the relationship between planning and urban policy, as planning can be identified as a political-administrative process of government, which, although it should be based on theoretical knowledge, needs to be defined as practical policies and guidelines (apud FERRARI, 1991).

The lack of good planning jeopardises people's quality of life and causes various social, environmental and ecological problems. For example, traffic jams in large cities are caused by a lack of planning, as shown in Figure 3. It is necessary to control the various activities and transformations that take place in municipalities in order to respect the limits of the environment.

Figure 3: Urban planning in São Paulo
Source: Mobilidade SP, 2011.

Well-developed urban planning within a city values environmental conservation and increases people's quality of life, ensuring their survival in the midst of big cities. According to Motta (2004, p. 25): "Planning is a form of learning. It is through the exercise of planning that we learn about external demands and needs and about the capacity of the municipal administration to respond. Even when not implemented, plans reveal the expectations and value references that are essential to a working group. People need references to keep up with contextual changes and the evolution of their own organisation."

"In order to properly manage urban drainage, knowledge of the area, its monitoring, the planning of [sic] actions aimed at minimising impacts and, above all, the participation and motivation of the population involved are indispensable." (COMITÊ PARDO, 2018) In addition to planning, another important factor is sanitation, since it directly influences the lives of the population. According to the WHO (World Health Organisation), sanitation is the control of all factors in man's physical environment that exert or can exert harmful effects on physical, mental and social well-being. In other words, it improves the quality of life of the inhabitants, contributing to the health of the population.

Also according to Cavinatto (1992), since ancient times man has learnt intuitively that water polluted by waste could transmit diseases. Civilisations such as the Greeks and Romans developed efficient water treatment and distribution techniques. Basic sanitation is therefore fundamental to preventing disease. In addition, keeping city streets and environments clean and avoiding accumulated solid waste prevents the proliferation of diseases and the appearance of animals such as rats and insects, which are responsible for spreading some of them. Knowing what sanitation is and its importance, it is important to have knowledge of its four aspects: the supply of drinking water, the correct treatment of sanitary sewage, urban cleaning and rainwater management.

3.3 Drinking water supply

Potable water is water that is suitable for human consumption and must meet drinking water standards. According to Barros et al. (1995), the Water Supply System represents the "set of works, equipment and services intended to supply drinking water to a community for domestic consumption, public services, industrial consumption and other uses".

3.4 Sewage treatment

The purpose of the sanitary sewerage system is to collect, transport and remove, treat and dispose of waste water in an appropriate manner from a sanitary and environmental point of view. The sewerage system exists to prevent the possibility of waste coming into contact with the environment.

with the population, water supplies, disease vectors and food.

3.5 Urban cleaning

Urban cleaning is not only associated with street cleaning, but also with all the maintenance of public cleanliness such as parks and street weeding. Rubbish is all the waste that comes from human activity. It must be well packaged to facilitate its removal. The better the final waste disposal systems are operated, the lower the impact on public health and the environment.

3.6 Rainwater management

Larentis (2016) explains that urban drainage is the practice used to dispose of rainwater in cities. With an efficient drainage system, problems with flooding are avoided in cities, which cause a great deal of damage to the affected region and the city as a whole. Each of these systems has its own characteristics that must be dealt with in a way that is compatible with the development of the municipality.

CHAPTER 4

Floods, waterlogging and inundation

□ Flooding is the overflow of water from a drainage channel into marginal areas. "The process of flooding occurs when the waters of rivers, streams and rainwater galleries leave the drainage bed due to the lack of transport capacity of one of these systems and occupy areas where the population uses for housing, transport, recreation, commerce, industry and others." (BARBOSA, 2006, p.33).

□ Flooding is the accumulation of water in the streets due to drainage problems. According to Grilo (1992): "Flooding generally occurs in flat areas or with depressions and valley bottoms, with surface runoff compromised by topography and the lack or insufficiency of a rainwater system in the urban environment. Furthermore, the smaller the green areas, the less water infiltrates into the soil, which feeds the suspended aquifers, causing less aid to surface run-off, which could mitigate the causes of flooding."

□ A flood is a large amount of surface runoff, i.e. when the water level rises without overflowing. It is caused by excessive rainfall or the discharge of an accumulated volume of water upstream, for example.

The Civil Defence of São Bernardo do Campo - SP (2015) defines flooding as the rise in the water level in the drainage channel due to increased flow, reaching the maximum level of the channel, but without overflowing.

Figure 4 below clearly defines and exemplifies how to differentiate between the three phenomena.

Figure 4: Waterlogging, flooding, inundation
Source: Folha de Ouro newspaper, 2017.

CHAPTER 5

Factors that increase the risk of flooding

"Water plays an important role in the urban environment, with the need to meet different demands, issues relating to its quality, availability and rainwater runoff. The management of this water is a large part of urban sanitation" (BRAGA, 2016, p. 7).

It is known that the first cities developed near watercourses because there people would have access to water, fertile land and could also use it as a means of transport.

As a result of urban and technological development, the growth of cities has led to investments to promote the sanitation of riverside areas, which has consequently led to the partial and, in some areas, total occupation of stretches of watercourse or areas reserved for natural flooding.

Unplanned urban occupation has led to drainage problems caused by high-intensity rainfall. Initially, the most affected areas were located close to watercourses. However, with territorial expansion and without adequate supervision of land use and occupation, the problems of flooding intensified and began to affect natural drainage points due to the city's planialtimetry and the degree of waterproofing.

A second factor that favours flooding is rubbish. All the rubbish that is dumped in the street during the first rains moves to the rainwater collection points and clogs these channels. In heavy rain, the water has nowhere to flow and accumulates. However, the issue of people throwing rubbish on the streets becomes a cultural process and not one of nature. Figures 5, 6 and 7 show cases of cities that have been hit by floods, São Lourenço do Sul-RS and Rio Branco-AC.

Figure 5 - Blocked manhole in Uberlândia - MG
Source: Correio de Uberlândia 2018.
Figure 6 Aerial view of São Lourenço do Sul - RS
Source: Schellin, 2011.

Figure 7. Aerial view of Rio Branco - AC
Source: G1, 2012.

According to a survey sponsored by the Ministry of National Integration, Brazil suffered more than 30,000 natural disasters between 1990 and 2012, giving an average of 1,363 events per year (BRASIL, 2013). Permeable areas are also factors that favour more frequent flooding.

Urban areas are almost entirely impermeable. Typical gardens and lawns are being replaced by pavements, and buildings have larger roofs, leading to an increase in the volume of water run-off.

When soil is waterproofed, it loses its ability to absorb water. Completely cemented land, pavements, asphalt and buildings make areas impermeable by forming a surface on the ground that prevents it from coming into contact with water.

When it rains, especially heavy rains, the soil is no longer able to absorb it, so the water looks for ways to run off and when it finds accessible points, it accumulates and begins to cause flooding.

Some municipalities have resorted to moving the riverbeds to another region, but in an attempt to get the water to drain quickly from the roads, the problem is displaced, making the investment more expensive. As a result, macro-drainage works have been started to provide a control solution for the main urban rivers, consisting of canal projects, floodplain embankment works and secondary works such as culverts, galleries and manholes; and micro-drainage, a solution to control the level of allotments, which includes the collection and removal of surface or underground water through galleries.

Other factors that aggravate flooding are:

- Inadequate project sizing;
- Inadequate works;
- High water table;
- Physical interference in the drainage system;
- Deforestation;
- Improper disposal of solid waste.

CHAPTER 6

Location of the avenues

Three avenues were selected for this study that are located in the lower areas of the city, Avenida João Paulo II, Avenida Dâmaso Drumond and Avenida Rosalvo Santos, as well as Avenida Dr Danilo Cunha, which connects these with the upper area of the city, as shown in Figures 8 and 9.

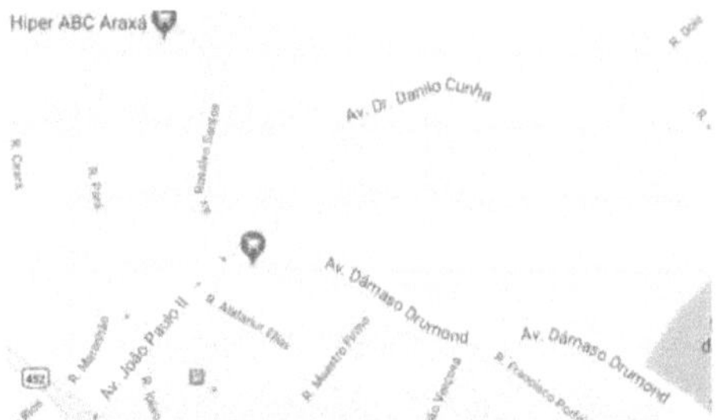

Figure 8. Avenues João Paulo II, Rosalvo Santos and Dâmaso Drumond
Source: Google Maps.

Figure 9. Urban location of the avenues selected for the study
Source: Google Maps.

6.1 Areas of influence

As they are located at the lowest levels in the city of Araxá, with approximately 2,250 metres at their lowest points, the avenues João Paulo II, Dâmaso Drumond and Rosalvo Santos receive surface runoff from the streets at higher levels and from the areas of influence that these streets aggregate. A list was made of the streets that have their runoff directed towards one of these avenues, as shown in Tables 1, 2 and 3, taking into account only the stretch that has a gradient facing them, directly or indirectly influencing the runoff from these lower streets.

VIEWS	LINEAR STRETCH LENGTH (METRES)
Avenida Vereador João Senna	1100
Abdanur Elias Street	450
Ipiao Street	570
Américo Autran Street	550
Imbiaça Street	500
Mário Campos Street	400
Maestro Firmo Street	290
Rua Bento Antônio	170
Geralcino Ribeiro da Silva Street	310
Canal Street	110
João Verçosa Street	350
Bom Jardim Street	170
John Paul II Avenue	1000
Damaso Drumond Avenue	800

Table 1. Extent of influence of the roads that flow into Avenida João Paulo II

VIEWS	LINEAR STRETCH LENGTH (METRES)
Rosalvo Santos Avenue	450
Avenida Amazonas	500
Slaughterhouse Street	400
Pará Street	450
Rio Grande do Sul Street	180
Ipiao Street	160
Ceará Street	110
Bahia Street	400
Alagoas Street	400
Travessa Pará Street	68
Rua Rio de Janeiro	350
Travessa Alagoas Street	55
Espírito Santo Street	93
Maranhão Street	170
Paraná Street	130

Table 2. Extent of influence of the roads that flow into Avenida Dâmaso Drumond

VIEWS	LINEAR LENGTH OF THE STRETCH (METRES)
Rosalvo Santos Avenue	450
Avenida Doutor Danilo Cunha	1700
Washington Barcelos Street	750
Rua Paulo Faria	240
Street F	54

Jason Armando de Paula Street	450
Cassiano de Paula Filho Street	350
Fernando Parolini Street	240
Brigido de Melho Filho Street	650
Maria das Graças Rosa Street	100
Rua José Nogueira da Fonseca	90
Ismar Ferreira da Silva Street	80
Three Street	75
Rua Maria Conceição Alexandre	70
Rua Vicente A. Barcelos	82

Table 3. Extent of influence of the roads that flow into Avenida Rosalvo Santos

Each of these roads was recorded and an approximate area of influence was calculated for each of the avenues. Area number 1 is the area that influences Avenida Damaso Drumond and Avenida João Paulo II concurrently as shown in Figure 10, Area 2 influences Avenida João Paulo II as shown in Figure 12, and Area 3 influences Avenida Rosalvo Santos as shown in Figure 14. To calculate these areas, sketches were developed in AutoCad that effectively approximated the real thing, as shown in Figures 11, 13 and 15, since there may be a difference, even if minimal, between the project and the real thing.

Figure 10. Area 1 - Area of influence of Avenida Dâmaso Drumond and João Paulo II
Source: Google Maps.

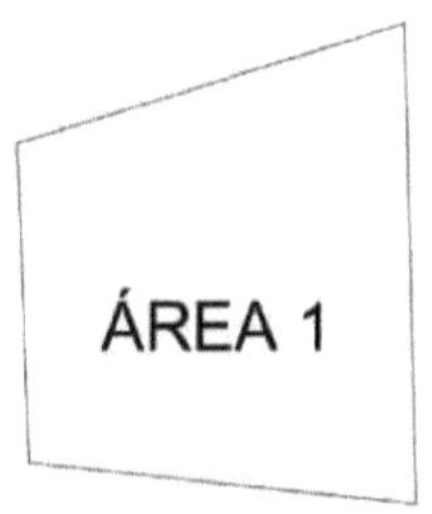

Figure 11. Sketch of area 1
Source: Prepared by the authors.

Figure 12. Area 2 - Area of influence of Avenida João Paulo II
Source: Google Maps.

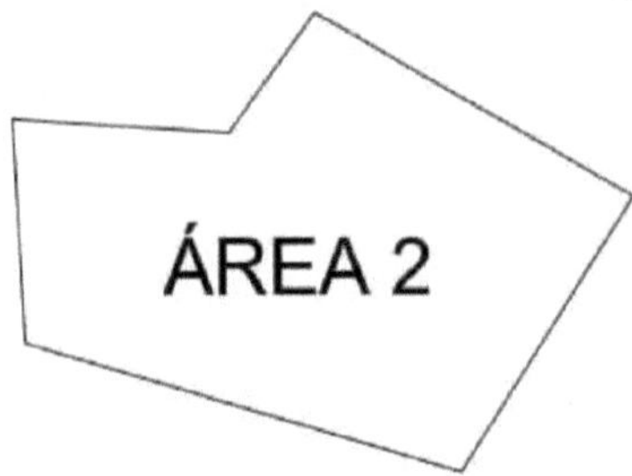

Figure 13. Sketch of area 2
Source: Prepared by the authors.

Figure 14. Area 3 - Area of influence of Avenida Rosalvo Santos
Source: Google Maps.

Figure 15. Sketch of area 3
Source: Prepared by the authors.

6.2 Accumulated rainfall

The monitoring and survey of the amount of rainwater accumulated in the last 3 years (2015, 2016 and 2017) was analysed in the city of Araxá. The National Institute of Meteorology (INMET) provides "reliable meteorological information to Brazilian society and constructively influences the decision-making process, contributing to the sustainable development of the country". (INMET, 2018). The annual accumulated rainfall graphs for 2015, 2016, 2017 and the first half of 2018 were analysed and archived, as shown in the graphs below.

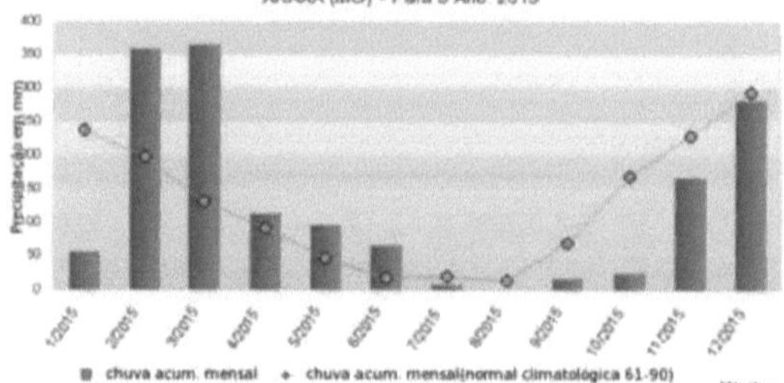

Graph 1. Accumulated monthly rainfall in 2015
Source: INMET.

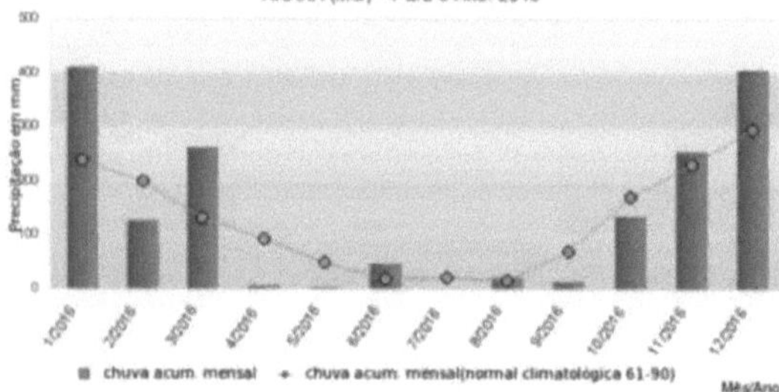

Graph 2, Accumulated monthly rainfall in 2016
Source: INMET.

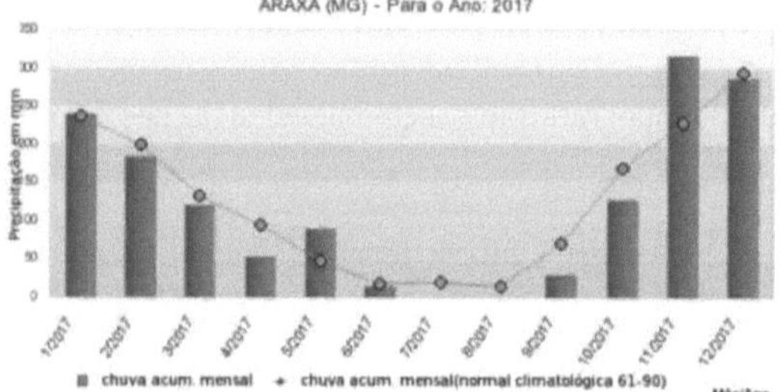

Graph 3. Accumulated monthly rainfall in 2016
Source: INMET.

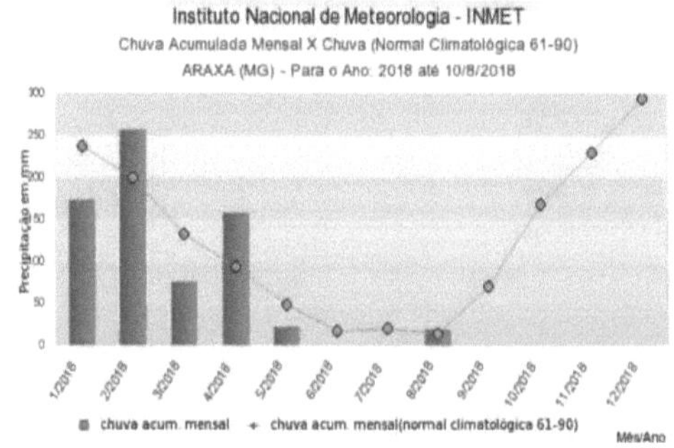

Instituto Nacional de Meteorologia - INMET
Chuva Acumulada Mensal X Chuva (Normal Climatológica 61-90)
ARAXA (MG) - Para o Ano: 2018 até 10/8/2018
Precipitação em mm
300
250
200
150
100
50
0
1/2018
2/2018
3/2018
4/2018
5/2018
6/2018
7/2018
8/2018
9/2018
10/2018
11/2018
12/2018
chuva acum. mensal
chuva acum. mensal(normal climatológica 61-90)
Mês/Ano

Graph 4. Accumulated monthly rainfall in 2018

Source: INMET.

Once the daily rainfall levels had been obtained, the approximate volume of water that would flow into the streets was calculated and the points where flooding had occurred and could occur were mapped. The data was tabulated in an Excel® spreadsheet and used to generate the mapping with the necessary data for each point identified and named.

6.3 Calculating areas

To calculate the areas of influence, a sketch of each of the three areas that were most similar to the real thing was designed in AutoCad and the size of these areas was calculated, as shown in the table below:

	APPROXIMATE AREA OF INFLUENCE
AREA 1	249890,56 M^2
AREA 2	202386,21 M^2
AREA 3	347716,43 M2

Table 4 - Approximate area of influence

6.4 Determination of study points

The drainage from the streets located in the higher areas to the lower streets in Araxá has been causing discomfort due to the history of flooding in certain areas of the avenues, caused by surface runoff and waterproofing caused by urbanisation. In order to choose the study points, historical surveys of flooding in previous years were carried out, as shown in the figures below.

Figure 16. Flooding in Araxá Source: Vale (2016).
Figure 17. Flooding in Araxá 02 Source: Vale (2016).

The three avenues, João Paulo II, Rosalvo Santos and Damaso Drumond, are interconnected by a roundabout and underneath them runs an underground gallery that rises to a higher level during heavy rainfall, making it one of the driving forces behind the flooding, as shown in the figures below.

Figure 18. Underground Gallery in Araxá Source: Ranieri (2014).
Figure 19. Underground gallery, following Avenida Rosalvo Santos Source: Ranieri (2014).

Figure 20. Rosalvo Santos Avenue Source: HZ Engenharia.
Figure 21. Following Rosalvo Santos Avenue Source: HZ Engenharia.

CHAPTER 7

Analysing floods

13 DECEMBER 2016.

One of the reasons for the flooding that occurred on 13 December 2016, as well as the flooded areas of later and/or earlier years, was the accumulation of water due to heavy rainfall (according to the Ciram classification (2016)), which had been occurring for five consecutive days. With levels above 16 mm/h, the rainfall on these days, considering days 9, 10, 11, 12 and 13, was one of the days with the highest levels of rainfall, second only to the previous day. Given that the rainfall on the 12th reached around 36 mm/h and that it lasted for almost 9 consecutive days, the underground gallery that receives the drainage of this water is supposed to have been saturated with this accumulation.

Other factors, such as urban rubbish, level problems, drainage and the saturation of the natural channel at the end of Avenida Rosalvo Santos, also influenced the occurrence of this and other floods, but the key factors were undoubtedly the abundance of water from the rains and the sizing of the gallery, which was not correctly estimated to sustain this abundance.

In order to base the amount of litres that rained on 5, 12, 13 and 14 December 2016, which were the wettest days of the month and all close to the day of the flood, we calculated the approximate amount of litres that rained on each of the days in the 3 areas chosen for this study. Bearing in mind that 100mm is equivalent to 0.1m, the results of this rainfall are shown in Tables 5 and 6.

	DAYS			
	5	12	13	14
MILLIMETRES OF RAIN	**80**	**33**	**29**	**59**
METRES	**0,08**	**0,033**	**0,029**	**0,059**

Table 5: Transformations of rainfall millimetres into metres

	AREA1	**AREA 2**	**AREA 3**
AREA (M)^2	249890,56	202386,21	347716,43
VOLUME OF RAIN (M^3) ON DAY 5	19991,2448	16190,8968	27817,3144
VOLUME OF RAIN (M^3) ON THE 12TH	8246,38848	6678,74493	11474,6422
VOLUME OF RAIN (M^3) ON THE 13TH	7246,82624	5869,20009	10083,7765
RAINFALL VOLUME (M^3) ON THE 14TH	14743,543	11940,7864	20515,2694

Table 6. Volume of rainfall on 5, 12, 13 and 14 December 2016

As you can see, the amount of litres of rainfall provided on these days, specifically in these selected areas, is of great proportions, so the drainage, run-off and absorption capacity should have an effective system to solve this problem efficiently and adequately.

Taking into account that all buildings respect the permeable area standards imposed by the IPDSA (the land must have at least 30 per cent permeable area), around 30 per cent of this rainfall would be absorbed and 70 per cent of it would be run off, clearly a very high proportion.

Nowadays, as these rules are not complied with, the percentage of surface runoff shown above increases significantly in the final amount, which helps to form these wetlands.

30 NOVEMBER 2018.

After almost 10 days of strong winds, extreme rainfall and/or very heavy rainfall (according to the Ciram classification (2016)), some trees fell, blocking a clandestine passageway that connects the urban area to a farm in the municipality. As a result of this obstruction, rainwater was trapped and flooded several streets and avenues, as shown in the pictures below.

The canal that runs along Damaso Drumond and Rosalvo Santos avenues overflowed due to the amount of water, lack of maintenance and cleaning of the canal, resulting in an even more aggravating flood, inundating a large part of the city's lower streets, damaging streets, invading houses and submerging cars.

IPDSA said that a new subdivision is being built on the site and these illegal crossings will be removed, creating new accesses to the subdivision.

Figure 22. Flooding R. Tonico Veloso
Source: Rogério Silva.
Figure 23. Flooding R. Tonico Veloso 02
Source: Rogério Silva.

Figure 24. Wilson Borges Avenue flooding
Source: Rogério Silva.
Figure 25. Flooding on Avenida Wilson Borges
Source: Rogério Silva.

Figure 26. Flooding R. Tonico Veloso 03

Source: Rogério Silva.
Figure 27. Flooding R. Tonico Veloso 04
Source: Rogério Silva.

Figure 28. Flooding Av.Wilson Borges 02
Source: Rogério Silva.
Figure 29. Flooding Av. Wilson Borges 03
Source: Rogério Silva.

Figure 30. Flooding R. Tonico Veloso 05
Source: Rogério Silva.
Figure 31. Flooding R. Tonico Veloso 06
Source: Rogério Silva.

Figure 32. Flooding R. Tonico Veloso 07
Source: Rogério Silva.

Figure 33. Flooding R. Tonico Veloso 08
Source: Rogério Silva.

Figure 34. Flooding R. Tonico Veloso 09
Source: Rogério Silva.

Figura 35.   Flooding Av. Wilson Borges 04
Source: Rogério Silva.

Figura 36.   Flooding Av. Wilson Borges 05
Source: Rogério Silva.

Figura 37.   Flooding Av. Wilson Borges 06
Source: Rogério Silva.

Figure 38. Congestion on Avenida Wilson Borges

Figure 39. Congestion on Avenida Wilson Borges 02
Source: Rogério Silva.

CHAPTER 8

Monitoring the daily and monthly backlog

Looking at the monthly accumulated rainfall graphs (Graphs 1, 2, 3 and 4), we can see that the months with the highest accumulated rainfall are: January, February, March, October, November and December.

According to Ciram (2016), hourly rainfall totals can be classified as: (i) extreme rainfall for values above 50 mm/h, (ii) very heavy rainfall for values between 16 and 50 mm/h, (iii) heavy rainfall between 4 and 16 mm/h, (iv) moderate rainfall between 1 and 4 mm/h and (v) light rainfall between 0.25 and 1 mm/h. Thus, the rainfall in January, February, March, October, November and December was classified according to the tables below.

PRECIPTATION IN MM7H	JANUARY (%)				FEBRUARY (%)				MARCH(%)			
	2015	2016	2017	2018	2015	2016	2017	2018	2015	2016	2017	2018
>50 (EXTREME RAIN)	0,00	0,00	0,00	0,00	16,66	0,00	7,14	0,00	9,52	5,56	0,00	0,00
16 to 50 (VERY HEAVY RAIN)	12,50	50,00	25,00	20,00	**33,33**	18,75	28,57	35,29	**33,33**	33,33	30,00	18,18
4 to 16 (heavy rain)	25,00	16,67	65,00	53,33	5,55	25,00	35,71	52,94	19,05	5,56	20,00	27,27
1 to 4 (MODERATE RAIN)	25,00	12,50	5,00	0,00	**33,33**	37,50	14,29	5,88	28,57	11,11	20,00	54,55
<1 (LIGHT RAIN)	37,50	20,83	5,00	26,66	11,11	18,75	14,29	5,88	9,52	44,44	30,00	0,00

Table 7. Rainfall classifications for the months of January to March.

PRECIPTATION IN MM7H	OCTOBER(%)			NOVEMBER (%)			DECEMBER (%)		
	2015	2016	2017	2015	2016	2017	2015	2016	2017
>50 (EXTREME RAIN)	0,00	0,00	0,00	0,00	6,25	13,33	0,00	11,11	10,53
16 to 50 (VERY HEAVY RAIN)	25,00	14,29	37,50	21,43	37,50	40,00	47,37	44,44	15,79
4 to 16 (heavy rain)	0,00	50,00	12,50	64,29	18,75	20,00	26,32	16,67	15,79
1 to 4 (MODERATE RAIN)	50,00	28,57	37,50	7,14	6,25	20,00	26,32	22,22	42,11
<1 (LIGHT RAIN)	25,00	7,14	12,50	7,14	31,25	6,67	0,00	5,56	15,79

Table 8. Rainfall classifications from October to December

In the general context of these months, it was observed that 3.81 per cent of the rainfall was extreme, 29.60 per cent was very heavy rainfall, 27.40 per cent was heavy rainfall, 23.23 per cent was moderate rainfall and 15.96 per cent was light rainfall.

CHAPTER 9

Mapping flooded areas

As a means of representing a total or partial surface of a space, the georeferencing carried out identifies the points surveyed as being the main sources of flooding, located on Dâmaso Drumond, João Paulo II and Rosalvo Santos avenues.

The table below shows the geographical positions of these points on each stretch of the aforementioned avenues.

EXCERPTS FROM AVENUES	JOHN PAUL II	DAMASO DRUMOND	ROSALVO SANTOS
GEOREFERENCING	19° 35' 06" S 46° 56' 49" W	19° 34' 57" S 46° 56 31"W	19° 34' 56" S 46° 56' 18" W
	19° 34' 59" S 46° 56' 34" W	19° 34' 58" S 46° 56' 26" W	19° 34' 54" S 46° 56' 34" W
		19° 35'02"S 46° 56' 19" W	19° 34' 50" S 46° 56'33" W
		19° 35' 06" S 46° 56' 06" W	19° 34' 47" S 46° 56'33" W
		19° 35' 18" S 46° 56'01" W	19° 34' 44" S 46° 56'35" W
		19° 35' 11" S 46° 56'01" W	19° 34' 44" S 46° 56' 36" W
			19° 34'45"S 46° 56' 36" W

Table 9. Georeferencing of the flood points on Dâmaso Drumond, João Paulo II and Rosalvo Santos Avenues.

CHAPTER 10

Sustainable street furniture

As discussed and highlighted in this issue, the main causes of flooding at the intersection of Avenida João Paulo II, Avenida Dr. Danilo Cunha, Avenida Dâmaso Drummond and Avenida Rosalvo Santos involve trees and rubbish inside the drainage galleries, causing obstructions, the widespread waterproofing of the city and the inefficient channelling of the Grande and Santa Rita streams.

The measures presented for reducing urban drainage problems are based on the use of sustainable systems that can be employed in the rainwater drainage system and that have quality and are methods that help to reduce the speed of surface runoff and prevent the dirt present in the streets from entering the manholes and culverts and reaching the galleries.

At the same time, the population must collaborate by avoiding littering and the city government must carry out the appropriate maintenance to ensure the efficiency of the elements suggested below.

There are various systems that can be used, such as infiltration trenches, rain gardens, green roofs and detention reservoirs, but these types of elements require larger spaces to be installed and the study site is already a completely urbanised environment.

Four models of street furniture used in the project in question will be presented.

11.1 Permeable paving

Permeable pavements are defined as those that have free spaces in their structure through which water can pass (FERGUSON, 2005).

As an alternative to the asphalt used, there is the possibility of installing blocks or interlocking paving. However, as the intersection of the avenues is a place that receives heavy traffic, both light and heavy, maintenance with any of these types of pavements would be high, which would make the project unfeasible.

So what was added to the project were drainage slabs for the pavements, since revitalising them is a quicker and more cost-effective process. The slabs are made of a porous concrete that allows water to drain away.

In Araxá, the IPDSA stipulates that houses must have at least 30 per cent permeability. Drainage slabs can also be used in this context, as they are an ecological floor.

Ceramic tiles have the following characteristics: they are athermic and slip-resistant, with high adhesion and resistance to friction, and they can be used as soon as they are finished.

When installing this type of flooring, it is essential that a specialised company carries out the laying so that criteria such as the water table and the soil's resistance capacity can be analysed. Below are illustrative images of drainage flooring.

Figure 40. Pavement drainage slab
Source: Rhino Pisos (2018).

Figure 41. Gutter drainage plate
Source: Rhino Pisos (2018).

Another use associated with drainage slabs is their adaptation to gutters. In this way, the same material can be used in different areas and the objective of reducing the speed of runoff will be achieved.

11.2 Filter strips

These filter strips, as shown in figure 36, are strips of vegetation that receive rainwater from surface runoff. This type of element also helps to retain solid materials, preventing them from reaching the manholes, reducing the speed of runoff and improving the water table, since the vegetation retains the water, favouring infiltration.

Figure 42. Filter strip
Source: Oliveira (2016)

11.3 Self-draining kerb

Currently on the market, there is a system called ACO Kerbdrain - shown in figure 37 - applied as a self-draining kerb. This device can be used in car parks and urban areas, i.e. areas that have the potential to accumulate water and have excellent hydraulic capacity.

Figure 43. Self-draining kerb
Source: External Works

11.4 Sustainable manhole

Kalena (2012) describes that sustainable urban furniture is already being used in São

Paulo, one of which is the manhole. A plastic basket is installed in which the rubbish is retained, as shown in the figure below:

Figure 44. Manhole, plastic basket
Source: Augusto (2012)

11.5 Retention basin

Retention basins are generally used to control water upstream, so that the amount of water reaches its destination slowly. In this case, these basins will be applied downstream, as the water will reach the galleries and, as it drains, it will be directed to these points; in this way, the water does not accumulate in the area.

Even with all this urban furniture, if the population doesn't receive environmental education about the

The problems have reoccurred. Maintenance and care are needed for everyone's well-being.

Appendix A shows the preliminary project before the urban furniture was installed. The project in appendix B then shows the use of the urban devices mentioned in this work, realised using Infra Works software. With this software it is possible to make an illustrative presentation that portrays the environment studied. It should be emphasised once again that this work is preliminary in nature and does not involve sizing or cost estimates.

CHAPTER 11

Project

The project was developed on the Infraworks platform, allowing the visualisation of basic parameters, planned structurally similar to the real thing as shown in the appendices.

Clearly, the design of the route created provides a defined and effective format for implementing this plan and sustainable urban furniture.

CHAPTER 12

Final considerations

Disorganised urban development poses numerous problems, leaving cities unprepared to deal with these impasses, causing serious negative impacts on the population and preventing continuous and credible growth.

Floods are often the result of the interaction between this disorderly development and the meteorological and hydrological phenomena that cause severe consequences.

The importance of noticing and localising inaccuracies in planning is the first step towards effective and continuous restructuring, which will bring about change and innovation. The floods in Araxá are caused by countless mistakes that are not reviewed and repaired.

Identifying the areas that influence surface runoff in low-lying areas and locating all the roads that form part of these areas was the first criterion, as the problem is not restricted to the designated points alone, but is contextual and interrelated with numerous factors and circumstances that contribute to these floods.

By linking the dates of the floods with the graphs of accumulated daily rainfall, it can be seen that the lack of dimensioning and drainage were significant in the urban planning of the areas chosen for the study, albeit in a negative way. Since surface runoff, the sealing of roads and land and the accumulation of water are not drained effectively and the canal that runs along Damaso Drumond and Rosalvo Santos Avenues is not sized to support these phenomena, these floods have become increasingly frequent and damaging.

The importance of noticing and localising inaccuracies in planning is the first step towards effective and continuous restructuring that will bring about change and innovation.

The floods in Araxá occur because of numerous mistakes - mentioned above - that are not reviewed and repaired. For these and other reasons, encouraging the use of sustainable urban furniture is essential for minimising these issues.

Therefore, only with proper planning will we be able to reduce the immobility of urban mobility, leading the country towards sustainable economic development.

CHAPTER 13

Bibliographical references

ALVES, João. Alagamento, Enchenteente ou Inundação? 2014. Disponívelem   : <http://aengenharianosensina.blogspot.com.br/2014/01   /foi-alagamento-enchente-ou-inundacao.html>. Accessed on: 26 December 2017.

ARCHITECTURE, Acr. Urban planning and its importance in the development of sustainable cities. 2017.   Available at   : <http://www.acr.arq.br/blog/importancia- planejamento-urbano>. Accessed on: 06 Mar. 2018.

AUGUSTO, Ronaldo de C.. Sustainable manhole is already being tested in São Paulo

BALDESSAR, Silvia Maria Nogueira. Green roof and its contribution to reducing the flow of rainwater run-off. 2012.

BARBOSA, Francisco de Assis dos Reis. Protection and control measures for urban flooding in the Mamanguape/PB river basin. 2006. Available at: <http://www.cprm.gov.br/publique/media/diss_francisc obarbosa.pdf>. Accessed on: 01 Dec. 2018.

BARRETO, A. B. L.; DIEB, M. A. Proposal for a linear park on the Jaguaribe River, between the neighbourhoods of Miramar and Cabo Branco, João Pessoa, PB. V!RUS, São Carlos, n. 14, 2017. [online] Accessed on: 08 March 2018.

BARRETO, Garcia. Architecture firm in AraxáMG   . Available at   :

<http://www.garciabarreto.com.   br/brasil/araxa/index.ht   ml#engenheiro-ambiental-sp-1>. Accessed on: 26 December 2018.

BOTELHO, Manoel Henrique Campos. Rainwater: Rainwater engineering in cities. São Paulo: Blucher, 2011. 297 p.

BRAGA, Júlia Oliveira. Waterlogging and flooding in urban areas: a case study in the city of Santa Maria -DF   .2016.   Available at: <http://bdm.unb.br/bitstream/10483/19267/1/2016_Juli aOliveiraBraga.pdf>. Accessed on:

13 Apr. 2018.

BRAZIL, Aco. Combined kerb and drainage system for modern and sustainable surface water managementCombined kerb and drainage system for modern and sustainable surface water management. 2016. Available at: <http: //www.acodrenagem.com.br/aco-kerbdrain/>.
Accessed on: 05 Mar. 2018.

CEAP. Microdrainage. Available   at:
<http://www.ceap.br/material/MAT28052014140255.p df>. Accessed on: 08 Nov. 2018.

CEAP. A brief history of urban planning in Brazil   . Available at:
<http: //www.ceap .br/material/MAT07082012164244.p df>. Accessed on: 13 April 2018.

CECIERJ. The urban phenomenon and the origins of urban planning. Available at: <http: //extensao.cecierj. edu.br/material_didatico/geo10 /html/01_02.html>. Accessed on: 08 March 2018.

Rain causes flooding in parts of Greater Vitória: Flooding points have already been recorded in streets in Vitória, Serra and Vila Velha. 2017. Available at: <https: //g 1.globo. com/e spirito - santo/noticia/chuva- causa-alagamentos-em-pontos-da-grande-vitoria.ghtml>. Accessed on: 23 December 2017.

Rainfall of around an hour leaves streets flooded in Araxá: Fire Brigade says water reached houses in São Francisco neighbourhood. In one hour it rained what is recorded in five hours, says station... 2015. Available at: <http://g1.globo.com/minas- gerais/triangulo-mineiro/noticia/2015/03/chuva-de-   aproximadamente-uma-hora-deixa-ruas-alagadas-em-araxa.html>. Accessed on: 11 March 2018.

CITIES. City Hall carries out emergency procedure on Avenida João Paulo II. 2017. Available   at:
<http: //www.diariodearaxa. com.br/prefeitura-realiza-   procedimento-emergencial-na-avenida-joao-paulo-ii/>.
Accessed on: 08 Mar. 2018.

CIRAM (2016). The November 2008 rains in Santa Catarina: a case study aimed at improving the monitoring and forecasting of extreme events. (Technical Note). Resource Information Centre

Santa Catarina Environmental and Hydrometeorological Centre. Available   at:
<http://www.ciram.com.br/ciram_arquivos/arquivos/gt c/dowloads/NotaTecnica_SC.pdf>.
Accessed on: 16 Feb. 2018

CLIMATE-DATA.ORG. Climate: Araxá. Available at   ·<https: //en. climate-data. org/location/24937/>.
Accessed on: 05 Feb. 2018.

COPASA, Cia de Saneamento de Minas Gerais. New PV buffer model. 2011. Available at:

<http: //www.copasa. com.br/wps/portal/internet/impren sa/noticias/releases/2011/junho/twitter-20110630- ie2054/!ut/p/a0/04_Sj9CPykssy0xPLMnMz0vMAfGjz OJ9DLwdPby9Dbz8gzzdDBy9g_zd_T2dgvx8zfULsh0 VAfwq3lw!/>. Accessed on: 13 Mar. 2018.

CORREIO, Redação. Rain causes property collapses and floods streets on Sunday (10): According to a Codesal bulletin released at around 11am, 81 incidents had been recorded up to that time. 2015.   Available at   : <https: //www. correio24horas .com. br/noticia/nid/chuva -causa-desabamentos-de-imoveis- e-deixa-ruas- alagadas-neste-domingo-10/>. Accessed on: 26 December 2017.

CPRM. HYDROLOGICAL   PROCESSES Floods, flash floods and waterlogging in the generation of risk areas. 2017. Available at: <https://defesacivil.es.gov.br/Media/defesacivil/Capaci   tacao/Materialial Didático/CBPRG - 2017/Processos Hidrológicos - Inundações, Enchenchentes, Enxurradas e Alagamentos na Geração de Áreas de Risco.pdf>. Accessed on: 25 Dec. 2017.

HIGHER EDUCATION COURSE IN WATER RESOURCE MANAGEMENT, 2005, São Paulo. Principios da Hidrologia Ambientaç. São Paulo: Partnership: Universidade Federal de Alagoas, Universidade Federal de Santa Catarina, @005. 203 p. Available at: <http://capacitacao.ana.gov.br/Lists/Editais_Anexos/At tachments/23/03.PHidrologiaAmb- GRH-220909.pdf>. Accessed on: 12 Feb. 2018.

DUARTE, Fábio. Urban Planning. Curitiba: Abdr, 2012. 199 p.

ENGINEERING, Hz. Sanitary Avenue - Araxá City Hall. Available at: <http://www.hzengenhariaeconstrucoes.com.br/ave   nida-sanitaria-prefeitura- municipal-de- araxa.php>. Accessed on: 15 Feb. 2018.

ENGINEERING, MC. Rainwater drainage. Available   at: <http: //mcengenhariabrasil .com.br/servicos-drenagem- de-aguas-pluviais.php>. Accessed on: 15 March 2018.

ESTADO, Agência. IBGE survey shows lack of greenery in cities. 2012. Available at: <http: //www.gazetadopovo .com. br/vida-e- cidadania/meio-ambiente/pesquisa-do-ibge- mostra-

lack-of-green-in-cities-2r3dw46qb94tot8fc665v1xla>. Accessed on: 09 Nov. 2017.

FERGUSON, B. K. Porous Pavements. Integrative Studies in Water Management and Land Development. Florida, 2005

GAETE, Constanza Martínez. Five principles of urban planning to make cities sustainable. 2015.   Available at   :
<https: //www. archdaily.com. br/br/770702/five-keys-of-urban-planning-to-create- sustainable-cities>. Accessed on: 13 Feb. 2018.

GIMENEZ, Alírio Brasil. Efficient rainwater drainage network can prevent flooding. 2018. Available at: <https://www. aecweb.com.br/cont/m/rev/efficient-pluvial-drainage-network-can-avoid-floods_10832_0_1>. Accessed on: 16 Nov. 2017.

GONÇALVES, Orestes Marraccini; OLIVEIRA, Lúcia Helena de. Building Rainwater Systems. 1998. Available at   :
<http: //www.pcc.usp. br/files/text/publications/TT_000 18.pdf>. Accessed on: 06 Nov. 2017.

Google Maps. Maps. Available at: https: //www.google.com.br/maps/place/Arax%C3%A 1 ,+MG/@-19.6039478,- 46.9731261,13z/data= !3m1!4b1!4m5!3m4!1 s0x94b03 701af9afc3 3: 0xf9d101560efb5abd!8m2!3d- 19.5906483!4d-46.9442412. Accessed on: 14 Feb. 2018.

HONDA, Sibila Corral de Arêa Leão. Environmental planning and urban land occupation in Presidente Prudente   (SP).2015. Available at:
<http: //www.scielo. br/pdf/urbe/v7n1/2175-3369-urbe- 7-1-0062.pdf>. Accessed on: 13 Feb. 2018.

IBGE.   Panorama.   2017.   Available   at: <https: //cidades.   ibge. gov.br/brasil/mg/araxa/panorama>. Accessed on: 09 March 2018.

INBS. The dangers of world population growth. Available at   :
<https://www.inbs.com.br/perigos-crescimento- populacao-mundial/>. Accessed on: 14 Apr. 2018.

INMET.   About   INMET.   Available   at:
<http://www.inmet.gov.br/portal/index.php?r=home/pa ge&page=about_inmet>. Accessed on: 10 May 2018.

Araxá Institute for Planning and Sustainable Development - IPDSA

NATIONAL METEOROLOGICAL INSTITUTE (Brazil). Accumulated rainfall in 24 hours: accumulated rainfall. 2017. Available at: http://www.inmet.gov.br/portal/ Data collected/____
INMET - National Institute of Meteorology .html>. Accessed on: 13 Feb. 2018.

URBAN FLOODING IN SOUTH AMERICA, 2003, Porto Alegre. URBAN FLOODING IN SOUTH AMERICA.Porto Alegre: Abrh, 2003. 157p   .   Available at   :
<file:///C:/Users/ilson/Downloads/INUNDACOES_
URBANAS_NA_AMERICA_DO_SUL (1).pdf>. Accessed on: 10 Feb. 2018.

LICCO, Eduardo Antonio; DOWELL, Silvia Ferreira Mac. Floods, Flash Floods and Flash Floods: Digressions on their socio-economic impacts and governance. 2015. Available at: <http://www.sp.   senac.br/blogs/revistainiciacao/wp-content/uploads/2015/12/110_IC_artigo-.pdf>. Accessed on: 01 Dec. 2017.

LOGICAMBIENTAL, Team. Why are there so many flooding hotspots in Macapá? 2016. Available at : <http://www.logicambiental.com.br/pontos-de- alagamento-macapa/>. Accessed on: 11 Dec. 2017.

LOURENÇO, Rossana. Sustainable Urban Drainage Systems. 2014. Available at: <http://files.isec.pt/DOCUMENTOS/SERVICOS/BIB   LIO/Teses/Tese_Mest_Rossana-Lourenco.pdf>.
Accessed on: 03 Mar. 2018.

LTDA, Travelbr Turismo. Minas Gerais. 2013. Available at: <http://www.cidades.com.br/cidades- do-brasil/estado-minas-gerais.html>. Accessed on: 09 March 2018.

MACÁRIO, Carol. Urban planning and environmental preservation are the way to avoid floods in Santa Catarina. 2017. Available at: <http://cbndiario.clicrbs.com.br/sc/noticia- aberta/planjamento-urbano-e-preservacao-ambiental- sao-o-cam caminho para evitar-as-enchentes-em-santa- catarina-198101.html>. Accessed on: 10 Dec. 2017.

MACEDO, Márcia Rejane; LIMA, Jéssica Helena de; MAIA, Maria Leonor Alves. Urban revitalisation and accessibility in the Recife district. Available at: <http://www.anpet. org.br/xxxanpet/site/anais_busca_o nline/documents/3_253_AC.pdf>. Accessed on: 23 March 2018.

MADEIRA, Maria Tereza Ribeiro. Urban furniture, a smarter city. 2013. Available at: <http: //www.arquitetare sponde .com.br/mobiliario- urbano-uma-cidade-mais-intelligente/>. Accessed on: 19 December 2017.

DISASTER MANUAL. Natural Disasters - vol. I. Ministry of National Integration, National Civil Defence Secretariat. Brasília-DF, 2003.

MARCHIONI, Mariana. Learn about sustainable drainage practices. 2017. Available at: <https: //www. construliga.com.br/blog/conheca-as- praticas-de-drenagem-sustentavel/>. Accessed on: 06 March 2018.

MARTINS, Mary. Even without heavy rain, streets of Vila Velha are flooded: Vila Velha City Hall reported that the municipality's Civil Defence was not called out on Tuesday (28) despite the rain that hit the city. 2017. Available at: <http://www.folhavitoria. com.br/geral/noticia/2017/03/   mesmo-sem-chuva-forte-ruas-de-vila-velha-   ficam-alagadas.html>. Accessed on: 09 Nov. 2018.

MIGUEL, Renato Abib Dutra; PEDROSO, Daiane Cristina; HOFFMANN, Rosa Cristina. The importance of urban planning and environmental management for the orderly growth of cities. 2009. Available at : <http://www.cronosquality.com/aulas/artigoplurb.pdf>. Accessed on: 09 Nov. 2017.

Ministry of Cities/Institute of Technological Research - IPT. (2007) Mapping risks on slopes and river banks. Brasília: Ministry of Cities/Institute for Technological Research - IPT.

MOTTA, C.P.C. Curso prático de direito administrativo. 2.ed. rev., atual. e ampl. Belo Horizonte: Del Rey, 2004.

MOURA, Erika Fernanda da Silva; SILVA, Simone Rosa da. Study of the degree of soil waterproofing and Proposals for Sustainable Urban Drainage Techniques in an area of Recife-PE. 2015. 16 f. Thesis (Doctorate) - Architecture and Urbanism Course,

National Journal of City Management, Recife, 2015

MUNICIPALITY OF CARMO DO PARANAÍBA (Org.). Various improvements have been made to public roads. 2015. Available at : <http: //www.carmodoparanaiba.mg.gov.br/noticias/17/ 06/2015/varias-melhorias-foram-realizadas-em-vias- publicas.html>. Accessed on: 18 Apr. 2018.

OLIVEIRA, Camila. 12 models of paving stones. 2016. Available at : <http://www.tudoconstrucao.com/12-modelos-de-para- calcadas/>. Accessed on: 12 Apr. 2018.

PAPASTAWRIDIS, Pedro. Urbanisation and public policies on urban mobility. 2013. Available at: <http: //www.administradores. com. br/article s/cotidiano/ urbanizacao-e-politicas-publicas-de-mobilidade- urbana/71498/>. Accessed on: 11 December 2018.

Paulo. 2012. Available at : <https://ambientalistasemrede.wordpress.com/2012/08/ 24/bueiro-sustentavel-ja-esta-em-teste-em-sao-paulo/>. Accessed on: 12 Apr. 2018.

PISOS, Rhino. Drainage board. 2018. Available at: <http://www.rhinopisos. com.br/site/produtos/18/placa_ drenante_piso_drenante_>. Accessed on: 13 Apr. 2018.

PISOS, Rhino. Products: drainage slab. 2018. Available at :

<http://www.rhinopisos. com.br/site/produtos/18/placa_ drenante_piso_drenante_>. Accessed on: 02 March 2018.

POMPÊO, Cesar Augusto. Sustainable urban drainage. Revista Brasileira de Recursos Hídricos, v. 5, n. 1, p. 15-23, 2000.

São Paulo City Hall. Crossing of Rua das Heras x Rua Mimosas has been completed: Vila Prudente Subprefecture finalises the reconstruction of the gutter in the stretch and makes the agreement, releasing the road. 2014. Available at: <http://www.prefeitura.sp.gov.br/cidade/secretarias/reg ionais/vila_prudente/noticias/?p=47846>. Accessed on: 13 Apr. 2018.

PROSAB. Urban rainwater management. Available at : <https://www. finep.gov.br/images/apoio-e- financiamento/historico-de-programas/prosab/prosab5_tema_4.pdf>. Accessed on: 13 Nov. 2017.

RANIERI, Caio. Couple on motorbike falls into canal on Avenida Rosalvo dos Santos, in Araxá. 2014. Available at   :
<http://www.minasnofoco.com/2014/08/casal-em- moto-cai-dentro-do-canal-na-avenida-rosalvo-dos- santos-em-araxa/>. Accessed on: 15 Feb. 2018.

REGION, Folha da. Rain floods the street and makes shopkeepers angry: Drainage channel clogged with leaves. 2018. Available at: <http: //www.folhadaregiao .com.br/região/chuva- faz- rua- ficar-alaga-e-deixar-comercias-irritados- 1.333000>. Accessed on: 11 Feb. 2017.

REIS JUNIOR, Ilson Rodrigues dos; GONÇALVES, Diogo Aristóteles Rodrigues; SILVA, Raiany Rodrigues. Flooding of low-lying municipal roads in the city of Araxá - MG. 2016. 30 f. TCC (Graduation) - Civil Engineering Course, UniaraxÁ, Araxá, 2016.

RIBEIRO, Júlia Werneck; ROOKE, Juliana Maria Scoralick. Basic sanitation and its relationship with the environment and public health. 2010. Available at: <http: //www.ufj f. br/analiseambiental/files/2009/11/TC C-SaneamentoeSaúde.pdf>. Accessed on: 13 December 2017.

SABOYA, Renato. The emergence of urban planning. 2008.   Available at : <http://urbanidades.arq.br/2008/03/o-surgimento-do- planejamento-urbano/>. Accessed on: 11 Feb. 2018.

SÃO BERNARDO DO CAMPO, Civil Defence. Flood, Floodway or Flood? Available at   :

http://dcsbcsp.blogspot.com.br/2011/06/enchente- inundacao-oralagamento.html Accessed May 2015.

SILVA, Adryely Julianne Silva da; FARIAS, Glorgia Barbosa de Lima de. Urban planning and sanitation: the causes of flooding in the city of Bragança-PA. 2017. Available at: <http://www.ibeas.org.br/congresso/Trabalhos2017/IX -007.pdf>. Accessed on: 03 Dec. 2017.

SILVA, Simone Rosa da. The impacts of urbanisation on urban drainage. 2015. Available at: <http://pec.poli.br/sistema/material_disciplina/fotos/S AU_Drenagem_urbana_Recife_2015.pdf>. Accessed on: 26 Nov. 2017.

SOUZA, C. Políticas públicas: uma revisão da literatura. Sociologias. 2002, n.16, pp. 20-45.

SUDERHSA. Drainage masterplan for the Iguaçu River basin in the Curitiba metropolitan region. 2002.   Available at   : <http: //www.aguasparana.   pr.gov.br/arquivos/File/pddr enagem/volume6/mdu_versao01.pdf>. Accessed on: 26 December 2017.

SUSTAINABLE, Ipdsa - Institute for Planning and Development. The Plan: Masterplan - A Development Strategy. 2018. Available at: <http://ipdsa.org.br/menu/link/7/o-plano>. Accessed on: 26 March 2018.

TASSINARI, Lucas Camargo da Silva. SIZING RAINWATER DRAINAGE SYSTEMS USING LOW IMPACT METHODS. 2014. 79 f TCC (Graduation) - Civil Engineering Course, Ufsm, Santa Maria, 2014.

Storm causes flooding in hospital and streets in the Cuiabá metropolitan region: Sectors of a university hospital in the capital were flooded and cars fell into a stream in Várzea Grande during the rain that fell on Friday (10).... 2017. Available at: <https://g 1. globo.com/mato-grosso/noticia/temporal- causa-alagamento-em-hospital-e-ruas-da-regiao- metropolitana-de-cuiaba.ghtml>. Accessed on: 26 November 2017.

TUCCI, Carlos EM et al. Urban flooding in South America. Ed. dos Autores, 2003.

TUCCI, Carlos. Urban drainage. 2003. Available at: <http://cienciaecultura.bvs.br/scielo.php?script=sci_artt   ext&pid=S0009-67252003000400020>. Accessed on: 27 January 2018.

TUCCI, Carlos. Urban rainwater management. 2005. Available at : <http://www.capacidades.gov.br/media/doc/acervo/069 06898a257ceb3ec8687675e9e36c8.pdf>. Accessed on: 13 Nov. 2017.

VALE, Ailton do. Rains punish Araxá and leave residents homeless: In a condominium, a torrent knocked down the wall of a house that fell on top of a gas cylinder; there was a risk of explosion and firefighters had to evacuate residents in a hurry. 2016. Available at: <file:///C:/Users/Ilson Junior/Desktop/Salvar/INICIAÇÃO 2017/RESUMO PARA MOSTRA/Chuvas castigate Araxá and leave residents homeless _ JORNAL O TEMPO.html>. Accessed on: 12 Feb. 2018.

VANESSA. Urban layouts - roads. 2017.
Available   at:
<https://arquiconcurseiros.wordpress.eom/2015/10/04/t racados-urbanos-vias/>. Accessed on: 14 March 2018.

VASCO, João Ricardo Justino. Sustainable Urban Drainage Systems. 2016. Available at: <https: //repositorio. ipl .pt/bitstream/10400.21/7168/1/D issertação.pdf>. Accessed on: 06 Apr. 2018.

VAZ, Valéria Borges. Urban drainage. 2014.
Available   at:
<http://www.comitepardo.com.br/boletins/2004/boleti m05-04.html>. Accessed on: 18 Dec. 2017

VI CONGRESO LBEROAMERICANO DE ESTUDIOS TERRITORIALES Y AMBIENTALES, 2014, São Paulo. URBAN SPRAWL AND THE CONSEQUENCES FOR DRAINAGE HEADWATERS: A CASE STUDY OF THE SOURCES OF THE VERTENTE 1 STREAM - UBERABA - MG. São Paulo: Estudio Territoriales, 2014. 1195 p. Available at: <http://6cieta.org/arquivos-anais/eixo4/Juliana Paula da Silva Rodrigues.pdf>. Accessed on: 12 Feb. 2018.

VIDAL, Fernando. Creation of the area of urbanisation in civil engineering courses in Brazil. 2017. Available at   :
<https: //www. univeritas .com/noticias/planning-urban-city-needs-and-architects-can-and-should-programme it>. Accessed on: 11 March 2018.

WEB, Aec. SUDS - Sustainable Urban Drainage System: A construction system that offers answers to common rainwater management problems in cities. 2018. Available at: <https://www.   aecweb.com.br/prod/e/suds-sistema-   urbano-de-drenagem-sustentavel_15120_29100>.
Accessed on: 05 Mar. 2018.

WORKS, External. ErbDrain drainage, 'Connecting Derby' transport scheme. 2017. Available at: <https: //www. externalworksindex.co.uk/entry/3077/A CO-Water-Management/KerbDrain-drainage- Connecting-Derby-transport-scheme/>. Accessed on: 12 March 2018.

CHAPTER 14

Appendices

APPENDIX A

Urban Planning Project - Before adding sustainable urban devices.

APPENDIX B

Urban Planning Project - Post-accession sustainable urban devices.

You can't teach anyone anything, but you can

help people discover it for themselves.

-Galileo Galilei

Printed by Books on Demand GmbH, Norderstedt / Germany